A Recipe for Prime Rib

Ed Zweiacher

Two Acre Publishing Company
El Reno, OK

ISBN: 978-0-578-52169-5

About the cover: The cover photo is from the Ft. Worth Stock Show and Rodeo. Pictured is the Brandon Horn family, Aven, Brek, Brandon and Jagger. Aven's steer was judged Grand Champion of the 2019 Steer Show by Judge Chris Mullinix. Bentley is a composite steer out of a Horn raised cow and sired by a Horn raised bull. He weighed 1400 pounds at the show and is ideal for what my book is about. I am anxious to see the carcass data on this steer. Grand Champions and Stock Shows are a major part of the Horn family and has been for generations. Winning is always nice and certainly nothing new to these hard-working people. When Brandon was a small boy, he won Grand Champion at Oklahoma City. M.E. Ensminger featured Brandon and his Shorthorn steer in his Feeds and Nutrition book. When I saw the photo of Aven, her steer and her family, I knew I needed that for my book. The more I thought about it and looked at it, the more I wanted it on the cover. Many thanks to the Ft Worth Stock Show and Rodeo for the photo and many thanks to Brek Binder Horn for giving me permission to use the family photo.

A Recipe for Prime Rib

Preface

I have come to love photography after taking a class and several workshops under Larry Clements. Because of this, I have been fortunate to be able to enter contests and fairs and win ribbons and publications such as calendars and so forth.

A thought came to me many times about developing a photo book that people could use on coffee tables and in doctors' offices. Then one day I saw an advertisement and an article in the El Reno Tribune about a course at the Canadian Technology Center titled, "Write, Publish, and Market Your Book" with Andrea K. Foster as the instructor. I enrolled with the goal of producing a photo book of my dreams.
The Book Lady, as Andrea is called, showed me the costs and returns of a photo book and the painstaking details of publishing one so I decided since I am not a quitter, I would just write a brief animal nutrition book, pass the class and be on my way.

As the course progressed, I heard more and more from the other students along with success stories and excitement. I found new friends and much help from the Book Lady and all her students and began to really enjoy the class. I began writing and decided along the way to not write a book on animal nutrition, as such but instead tackle the task of writing some information for acreage or small farm owners on how they could raise their own beef.

I also enjoy good freezer beef myself and I am very fond of prime rib so I decided to title my first book "A Recipe for Prime Rib" and discuss a possible way to raise a calf for your own Prime Beef. The first course finished very rapidly so I enrolled in the next class and then the next class and again. After 4 eight-week sessions over two semesters I am almost finished with my first book and already planning two more, one of which will be a photo book in spite of it all.

I thought about listing all the people who have had an influence on my life, career and this book. As I started out with the abovementioned people along with Dr. Ted Montgomery, Dr. Ralph Durham, Dr. Jim Eischen, Dr. M.E. Ensminger, Ervin Emmert, Bob Hooker, Jeffrey Reuter, Cotton Furnish, Milton England, Dr.Jerry Martin, Dr. Estes Firestone, Dr. Robert Albin and many more, I concluded there are so many people who have helped me in many ways before and during my career that it would be impossible to name them all here.

INTRODUCTION

Even though there are many Beef Lovers in the United States, we still rank fourth in per capita consumption. We are 40 pounds below the Number One beef eating country, Uruguay. That should not deter our quest for quality over quantity.

Quality in beef means tenderness, flavor, color and leanness. Mainly it means customer satisfaction. That is what each producer of beef should strive for, and hopefully, this book will help you do just that. A good friend told me today "You just can't buy meat in the grocery store like we used to". He was referring to the quality of the meat today. When you raise your own beef, you will be in control of the quality.

Many people would like to attain a suburban to rural lifestyle, whether on the edge of a city or out in the boondocks. In addition, these types of people desire healthy, wholesome foods they produce themselves. This not only gives them assurance of a quality product produced with or without chemical pesticides, they also receive satisfaction in knowing they produced the food with their own hands and in their own way. They know what they are eating and what they are feeding their children.

When purchasing food items from any other source, there is always that measure of doubt as to where the food is from and who produced it with what methods. If you produce some of your own food, you eliminate a bit of that doubt.

It is my hope that this book will give you some ideas on managing your own beef production from year to year. The book covers the acquisition of cattle and the facilities you will need to handle the animals. Feeding and good nutritional practices as well as calf health will be explained. Ideal management practices and ideas for your specific production will be presented. The book will give you some goals to shoot for such as ideal weights before processing and after harvest weights that you can expect. Also, you will learn some new terms and revisit old terms including "buzzwords" and phrases used in the industry. Even though the book is mostly instructional, I will throw in a story or two to illustrate an idea or just to make it lighter reading.

"Never let your steer have a bad day" is a phrase I learned at a seminar at Express Ranches in Yukon, OK one evening from a professor at Texas A&M University. That comment has stayed with me since, because it means a lot. Yes, to make sure your calf never has a bad day, ever, would be very difficult and very expensive. However, you can make this phrase a goal to work toward. You will read this phrase throughout the book. I will also explain why I used the term steer in that phrase.

Many of you will want to grow a "grass fed" beef even if you have never eaten any "grass fed" beef. This may be because you have heard from someone how healthy or lean grass fed can be or maybe because it sounds good. I do not want to discourage you from such an endeavor, but I will ask "why?" Fat provides flavor in meat. I am not talking about external fat but the fat within the muscle called marbling, the white flecks of fat scattered throughout the lean itself. In addition, the highest quality of beef is Prime, and the only way to get close to that quality is by feeding high-energy diets that result in marbling or stored energy in your final product. You should not be afraid of the term fat. It means finished or ripe and ready for harvest.

Kylie and Cambrie Callis with their new calf. You can bet these cute girls will name their new calf for sure. If you have children who want to name the calves, it might be wise to feed heifers so you could keep one for breeding and not the freezer.

Probably one of the most important things I should have written in the first or second paragraph is, "DON'T NAME YOUR STEER." A good friend of mine said one time that the easiest way to start losing money on cattle is to name your cows. If you name your animals, you will lose money. Well, with my ideas in this book, you are not going to profit from this production dollar - wise. Your profit will be in the knowledge that you grew your own beef and the quality it turned out to be. There is another reason you should not name your steer, and that is because you will be eating the product. Your steer is not a pet, but we hope he acts like one, so we can make sure he never has a bad day.

Well, I just visited with a member of the Oklahoma Swiss Club and told him about my book. I also told him that I strongly suggested the readers not to name their calves. He said that is strange, because, in Switzerland, all the cows have names. This is also true for the ones shown in livestock shows and tournaments in the United States. Most of the animals with names are used for milk, cheese, or breeding purposes and only for meat when they complete their cycle of other production. If you name your animals, that is your business. I just want you to be able to enjoy some good beef.

At the conclusion of my classes, I normally ask if there are any questions or comments. I want to do the same at the conclusion of this book, giving you my contact information in various forms. I do hope you enjoy the book and gain from it. My goal is to help you produce an enjoyable, delicious beef that is of quality, healthy and wholesome.

Enjoy your beef.

CONTENTS

Chapter One
Facilities and Preparations Before Purchasing...13

Chapter Two
Selecting your Calf for Best Beef Growing Potential... 21

Chapter Three
Purchasing and Transporting... 35

Chapter Four
Handling Cattle...41

Chapter Five
Feeding and Nutrition ... 45

Chapter Six
Health and General Care... 53

Chapter Seven
Processing to Freezer Beef...63

Chapter Eight
Let's Talk Beef... 67

Chapter Nine
Costs and Returns... 73

Check List... 75

Cowman Terminology... 77

Notes

CHAPTER 1

Facilities and Preparations Before Purchasing

A good fence starts with a good corner post, and good fences make good neighbors. You want to make sure you have a good fence, strong, stout and high enough for cattle before you unload anything on your property. Always make sure your gates are closed when they need to be.

Build your fence for cattle, not other species. You can look on YouTube or just Google good cattle fences. A pasture fence can be different from a pen fence too. I have seen folks try to make an "economically feasible" fence or pen. That did not work for the intended purpose. Well, I have even seen some try to use just a rope as a fence for cattle. That does not work, either. If your steers are very docile and move slow at every turn, perhaps that method might work, but the truth is, they may not be gentle and slow at first, especially with new surroundings and new people. You need to be prepared to accept their plight or even flight on the first arrival. A good solid fence will hold them until they get used to you and your place.

Do not be afraid of a few gates in your fence and especially pen area. If you have a gate on your perimeter fence, please keep it locked. A perimeter gate is just in case a neighbor animal gets in your pasture you can get them back easily. Hopefully, your fences will keep your calves in. Gates can be useful in moving and sorting calves. It is also possible to squeeze a calf behind a gate or between a gate and a solid fence. This might give you an opportunity to do some minor work on the calf like spray or pour on. Squeezing behind a gate is not useful when checking for ear ticks or placing an ear tag in a calf. You will need a head gate for that. If you have an acreage and grass or planted crops to graze, this will be a plus.

The stocking rate for cattle varies according to location, size of animal, and the crop growing. If you graze your calves, you will have to make the determination as to how large a pasture you will need. Watch your grass, and do not let the pasture be grazed to the nub. Grass can recover better if it has a good root system, and the amount of above ground grass is indicative of the amount of root system it has. Grazing young steers can get some growth on them before the fattening process should begin. You can get the skeletal and muscle growth in a dry lot situation but gaining skeletal and muscle tissue works best if you have a bit of pasture. For two calves, you will probably need between 2 and 5 acres. You can also feed your high-energy diets while the cattle graze. This works ideally if you are seeking a "natural fed" beef.

Electric fences work and are just fine, if built correctly. I know of a 900-acre ranch in Oklahoma that uses only electric fences for the entire ranch

Nice crossbred steers on winter wheat behind a nice 5–wire fence.

pasture fences, not the pen fences. Many show cattle operations use electric fences to keep the animals from rubbing hair off. Before you use electric fences, be sure to study up on the subject. You need to know how they work and when they work. If the electricity quits for any reason, you may as well have no fence at all and may be looking for your cattle for a long time. In addition, you have to keep in mind, humans may accidently walk into the electric fence or reach out to grab it, not knowing what will happen. I saw a deal on Facebook once that said, "Some people learn by listening, some learn by watching, and others just have to pee on the electric fence for themselves." I like some electric fencing, and it works well. Just be careful and knowledgeable about the fencing. My recommendation might be to use it inside your perimeter fencing only. You will need to train your animals to electric fencing.

Pen space also varies according to your location, drainage and slope. To have a pen that has a solid base of something similar to caliche is ideal. I do not recommend concrete or slats, but some people may prefer those. The soil you have now may be sufficient, but remember, cattle will make a mud hole just from their hoof traffic over time. If you get a lot of annual rainfall, please adjust accordingly for your pen space. There are some recommendations of 300 to 800 square feet of space per animals in feedlot pens. If you are only using your pen space to catch and work (vaccinate, doctor or deworm), you can use a bit smaller space. The main thing about space for your animals is to make sure they never have a bad day. Keep them dry, comfortable and content.

Shade is important for most areas, especially in summer. Provide enough space under shade for all your animals, as well as a windbreak or more protection in winter from harsh winds and moisture. Remember, moisture normally flies with the wind, but when it hits a windbreak, it falls. Do not force your animals to stand where snow might fall and cause them to breathe the snow or moisture of any kind. Doing so could result in drowning and be a very expensive lesson. Consider your wind direction, amount of rain per year, and all environmental factors when building or remodeling pens. This will help you decide if you need just a shed or a barn. If you decide to use a barn and small pen to get animals shelter, be prepared to clean the manure each day. Daily cleaning will be easier than waiting to clean on a weekly or monthly basis. Cleaning will reduce the fly

Cambrie Callis working with her dad in the barn. Please notice the fence. It is high enough so they cannot jump it, stout enough they cannot go through it and the spacing is just right for heads to not get caught in between. The barn is built to get the calves out of the weather and still allow a breeze in summer and avoid condensation in the winter. The fence is a pen fence and not for pastures. There is an alley in front and gates leading in and out of the alley. You do not have to tie your cattle like Brandon Callis is doing but be around them enough to gentle them down.

issue as well.

Feed bunk space requirements seem to be about 22 to 26 inches per steer. A normal feed bunk would be 18 inches in width and for two steers, four feet in length. This would be open on one side only and could have a hayrack above, if you plan to hand feed hay. An ideal situation would be a feed bunk inside a stall in the barn, where you could feed grain and hay in same bunk. A round bale could be set inside close to the feed bunk and a feed bin or box that you could use a pitchfork for the hay and scoop from the feed bin or box for the mixed ration. All this would be handy for you, and keep the steers clean, dry and out of the weather for the feeding time. This would also aid in making them more docile, or gentle them down a bunch, as we might say.
Good clean water is very important, and the water location is important as well. If you have a 2-acre or 5-acre pasture and you want the calves to enter a pen easily, you might consider placing your water source inside the pen. The calves will come inside to get water at least twice per day, and when they do, you can just shut the gate on them and have them in the pen. When they first arrive, place them in the pen with water, so they can find it easily, and they will remember where it is. They must always be able to access the water. They can live without feed longer than they can live without water, so check your water every day when you feed. You do not want them to

go without either water or feed (and grass is feed, too); otherwise, they will have a bad day, and you do not want that.

Safety for your animal and you must be a consideration when designing your pens, fences and feed bunk areas. Sharp corners and sharp protruding objects can cause serious bruising and cuts, even to a strong hided animal and especially humans. Look around your space where animals are kept and keep it free of debris and harmful objects. Remember, the new calves may be a bit skittish and jump at various stimuli. Learn quickly what your animals may or may not do when you approach and know how to approach them with least confusion to the animal. You do not want to be the cause of an injury. Be careful and be patient. Move slowly around the animals even when you have all the facilities in safe order.

My mother used to say, "Cleanliness is next to Godliness." So, if you want to be completely devoted to making this the best prime beef you possibly can, keep your pen clean. When you design your feed pen or catch pen, look for ways to make the cleaning part easier. Think, "Can I get the wheel barrow through there?" "Can I get the Kubota with a hydraulic dump in this gate?" "Will that be easy to scoop from that surface?" and many other questions and ideas about keeping the area clean. Once again, doing the cleaning daily is much easier than cleaning once a week or month. There may be one exception to that statement, and that would be for pens outside which you could scrape with a blade on your tractor. One idea may be to feed in the pasture as much as you can. This could be from a desire to have natural beef as well as not having to clean as often. Just remember to get the calves inside or under shelter on bad weather days.

Below are a few URL's to get you started looking at pens, barns and fence designs.
https://www.youtube.com/watch?v=ssT0n_sKn70
https://greenhorns.org/2012/02/23/university-of-tennesseeext-ag-building-plans/ https://thefreerangelife.com/farm-fence-options/

Leonard and Joe Hale barn in El Reno, Oklahoma. This is an older barn and much larger than you would need on an acreage. However, if you had a barn one fifth this size and fences like you see here, the facility would fit your need for feeding two steers.

Chapter 2

Selecting your Calf for Best Beef Growing Potential

Two calves on no-till wheat pasture. The smaller one is a stocker calf and the larger one would be called a feeder. There is 6 or 7 months difference in age.

The first thing that needs to be said here is, "When you purchase one calf to raise, purchase another one, too." Keep the calves in pairs, because they like a "buddy." This may be beneficial with competition at the feed bunk, resulting in faster gains and more growth. Buddies give a sense of security to the animal, makes them feel more comfortable. Dr. Fred Reuter, a veterinarian near El Reno, OK, told my students to always keep your herd in even numbers. They have a buddy system. Another plus to this might be that you can sell the second steer for a premium and help offset some of the total costs. Keep in mind, this may not be a profitable enterprise. Your reward will be in excellent beef and pride.

When the King Ranch was at its prime, they only wanted red cattle (Santa Gertrudis and Santa Cruz) and red "sorrel" horses. Some people like the color of the Belted Galloway, while others prefer Charolais. The breeds of cattle, and there are many, can be reviewed in photos with some typed information by googling "Breeds of Cattle." What you will find there is a website from Oklahoma State University on Breeds of Livestock. Researching this website can give you information on the various breeds. One thing you should consider is that each breed is limited to the genetics within that breed. If you like the rib eye area (square inches of muscle at the 12th rib or size of steak)of a Limousin, you have to choose Limousin. If you like the marbling of an Angus, you must choose an Angus. If you like both, perhaps you need a Limousin Angus cross. By selecting the two breeds, you will also reduce the effect of each gene you are going after. As an example, the Limousin will reduce the effect of the marbling in the cross.

Angus will marble better than any other beef breed, so you really need Angus in the genetics of your beef. Today, the Certified Angus Beef program is going strong because of promotion and a Quality product. Breed and color are your choice; I am just giving my ideas and thoughts. The Texas Cattle Feeders Association, Amarillo, TX, host a live animal and carcass contest for feedlots. Any feedlot wishing to compete can pay an entry fee and bring a specified number of cattle put together for uniformity of Quality and Yield and be judged live and on the rail. More often than not, a pen of Charolais Angus cross steers will win the contest. The Charolais genetics give the harvested product high cutability or good Yield Grade, and the Angus genetics give the carcass plenty of marbling for a high-Quality Grade. The combination results in an excellent beef, often. If you tour any feedlot in the Texas and Oklahoma Panhandles, you will see 90% or more crossbred cattle in the pens. They may not be Charolais Angus cross but are of some cross. The reason for this is hybrid vigor found in crossbreeding. Anyway, after reviewing the breeds of cattle, perhaps you may prefer a certain breed or crossbred. My personal preference would be a Charolais Angus cross, a mouse colored calf with a black not, much like the steer on the cover.

There is currently a great deal of promotion for the Waygu breed of cattle because of their enormous amount of marbling. If you love fat and plenty of it, you may consider an Angus Waygu cross.
Maturity refers to the physiological age of the carcass. Maturity is determined by a USDA Grader. Marbling and maturity are used to determine the Quality Grade of the product. You can control the maturity by harvesting at an age between 14 months

and 24 months of age. To know the birth dates of your calves or at least a birth month would be good. Most ranches calve in spring and/ or fall, but some farmers may calve year-round. If you learn your calves are spring born, then that gives you an idea of the month, give or take a couple. The BQA (Beef Quality Assurance Program) says calves need to be harvested before 30 months of age. If you buy six to eight-month-old calves, run them on grass for 4 months, and then feed them 6 to 8 months, your cattle should be harvested at 18 to 22 months of age. Keep in mind, younger is better, but you want the right degree of finish on them, too.

This is an Angus/Charolais cross steer. With snowflakes the size you see here, he may not be having a good day. If they have a good amount of finish, they can withstand some cold temperatures.

Bradon Callis with his Champion Heifer at the SW District show in Oklahoma. She has some Angus and Charolais genetics along with Simmental. She is called a Composite female. Bradon will keep his heifer for breeding purposes but for this book, she is a great picture of what Prime needs to look like. She is a good one.

Back fat measured at the twelfth rib and rib eye area are the major factors in determining Yield Grade on cattle. There are a couple of other factors, but they do minimal change to a yield grade. Yield Grades tell us the amount of muscle as opposed to amount of fat on the carcass. The ability to marble, put on external fat, and the size of the rib eye area are genetic traits.

When you go to purchase your cattle, you want to start with animals that have good carcass traits. If you purchase from a ranch or farmer, they will most likely know if their cattle possess these traits. Back fat will be determined more by how you feed your calves than the genetic trait associated with it. You want to strive for ½ of an inch of back fat. This amount seems to make the cattle grade at least choice in most cases. You may go a bit past that to reach prime.

Temperament may throw a kink in some of your decisions on breed. You really should try to select for docility, but that may not happen the way you would like it to. Some exotic breeds seem to have a bad or not so good temperament as compared to others. If you travel to Europe and look at the handling of cattle there, you may wonder why we in the United States have trouble with some of the European breeds. Perhaps they sold us their culls, according to docility. I am not sure. There are rogues in all breeds, but for the most part, Angus and Hereford seem to be gentle. If you have the solid and good set of fences and pens, you can gentle your calves down with regular feedings and care. The first place to start is for you to be gentle to them as best you can.

Bulls gain faster than steers, and steers gain faster than heifers. On the other hand, heifers fatten faster than steers, and steers fatten faster than bulls. Thinking about this as a combination, which sex would be preferred? Yes, the preference should be steers, with heifers as a second choice. Your beef product will be finished at less than 24 months of age, but that is still time enough to get too much testosterone going with an intact male, which would have a negative effect on the quality of beef. Also, the heifer would begin to have a high level of estrogens in her system by that time and would exhibit signs of estrous on a regular basis, which could affect gains or growth. A castrated male bovine, a steer, is your best choice for your beef. Next, we will talk about purchasing and transporting your Charolais Angus steer, just sayin'.

Here are the USDA (United States Department of Agriculture) standards for grading feeder cattle.

Feeder Cattle Grades and Standards

Grades of Thrifty Feeder Cattle (Frame Size)

1. **Large Frame (L)**. Feeder cattle which possess typical minimum qualifications for this grade are thrifty, have large frames, and are tall and long bodied for their age. Steers and heifers would not be expected to produce U.S. Choice carcasses (about 0.50 inch fat at twelfth rib) until their live weights exceed 1250 pounds and 1150 pounds, respectively.

2. **Medium Frame (M).** Feeder cattle which possess typical minimum qualifications for this grade are thrifty, have slightly large frames, and are slightly tall and slightly long bodied for their age. Steers and heifers would be expected to produce U.S. Choice carcasses (about 0.50

inch fat at twelfth rib) at live weights of 1100 to 1250 pounds and 1000 to 1150 pounds, respectively.

3. **Small Frame (S)**. Feeder cattle included in this grade are thrifty, have small frames, and shorter bodied and not as tall as specified as the minimum for the Medium Frame grade. Steers and heifers would be expected to produce U.S. Choice carcasses (about 0.50 inch fat at twelfth rib) at live weights of less than 1100 pounds and 1000 pounds, respectively.

Grades of Thrifty Feeder Cattle (Thickness)

4. **No. 1.** Feeder cattle which possess minimum qualifications for this grade usually display predominate beef breeding. They must be thrifty and moderately thick throughout. They are moderately thick and full in the forearm and gaskin, showing a rounded appearance through the back and loin with moderate width between the legs, both front and rear. Cattle show this thickness with a slightly thin covering of fat; however, cattle eligible for this grade may carry varying degrees of fat.

5. **No. 2.** Feeder cattle which possess minimum qualifications for this grade usually show a high proportion of beef breeding and slight dairy breeding may be detected. They must be thrifty and tend to be slightly thick throughout.

They tend to be slightly thick and full in the forearm and gaskin, also showing a rounded appearance through the back and loin with slight width between the legs, both front and rear. Cattle show this thickness with a slightly thin covering of fat; however, cattle eligible for this grade may carry varying degrees of fat.

6. **No. 3**. Feeder cattle which possess minimum qualifications   for this grade are thrifty and thin through the forequarter and the middle part of the rounds. The forearm and gaskin are thin and the back and loin have a sunken appearance. The legs are set close together, both front and rear. Cattle show this narrowness with a slightly thin covering of fat; however, cattle eligible for this grade may carry varying degrees of fat.

7. **No. 4.** Feeder cattle included in this grade are thrifty animals which have less thickness than the minimum requirements specified for the No.3 grade.

I want to add the USDA standards for quality grades for your finished or fattened steers also:

1. **Prime**. Slaughter steers and heifers 30 to 42 months of   age possessing the minimum qualifications for Prime have a fat covering over the crops, back, ribs, loin, and rump that tends to be thick. The brisket, flanks, and cod or udder appear full and distended and the muscling is very firm. The fat covering tends to be smooth with only

slight indications of patchiness. Steers and heifers under 30 months of age have a moderately thick but smooth covering of fat which extends over the back, ribs, loin, and rump. The brisket, flanks, and cod or udder show a marked fullness and the muscling is firm.

> a. Cattle qualifying for the minimum of the Prime grade will differ considerably in cutability because of varying combinations of muscling and degree of fatness.

> b. Cows are not eligible for the Prime grade.

2. **Choice.** Slaughter steers, heifers, and cows 30 to 42 months of age possessing the minimum qualifications for Choice have a fat covering over the crops, back, loin, rump, and ribs that tends to be moderately thick. The brisket, flanks, and cod or udder show a marked fullness and the muscling is firm. Cattle under 30 months of age carry a slightly thick fat covering over the top. The brisket, flanks, and cod or udder appear moderately full and the muscling is moderately firm.

> a. Cattle qualifying for the minimum of the Choice grade will differ considerably in cutability because of varying combinations of muscling and degree of fatness.

3. **Select.** The Select grade is limited to steers, heifers, and cows with a maximum age limitation of approximately 30 months. Slaughter cattle possessing the minimum qualifications for Select have a thin fat covering which is largely restricted to the back and loin. The brisket, flanks, twist,

and cod or udder are slightly full and the muscling is slightly firm.

    1. Cattle qualifying for the minimum of the Select grade will differ considerably in cutability because of varying combinations of muscling and degree of fatness.

4. **Standard.** Slaughter steers, heifers, and cows 30 to 42 months of age possessing the minimum qualifications for Standard have a fat covering primarily over the back, loin, and ribs which tends to be very thin. Cattle under 30 months of age have a very thin covering of fat which is largely restricted to the back, loin, and upper ribs.

    a. Cattle qualifying for the minimum of this grade vary relatively little in their degree of fatness. Therefore, the range in cutability among cattle that qualify for this grade is somewhat less than in the higher grades.

(from the United States Department of Agriculture Marketing Service, updated 2002)

GUIDE TO EVALUATION OF SLAUGHTER CATTLE

U.S. SLAUGHTER STEER GRADES

U.S. YIELD GRADES   **U.S. QUALITY GRADES**

ASWeb-121

Yield Grade 1 — Prime

Yield Grade 2 — Choice

Yield Grade 3 — Select

Yield Grade 4 — Standard

Yield Grade 5 — Utility

Photos: USDA Agricultural Marketing Service, July 2001

USDA Grades

The USDA grade shields are highly regarded as symbols of safe, high-quality American beef. Quality grades are widely used as a "language" within the beef industry, making business transactions easier and providing a vital link to support rural America. Consumers, as well as those involved in the marketing of agricultural products, benefit from the greater efficiency permitted by the availability and application of grade standards.

Beef is evaluated by highly-skilled USDA meat graders using a subjective characteristic assessment process and electronic instruments to measure meat characteristics. These characteristics follow the official grade standards developed, maintained and interpreted by the USDA's Agricultural Marketing Service.

Beef is graded in two ways: quality grades for tenderness, juiciness and flavor; and yield grades for the amount of usable lean meat on the carcass. From a consumer standpoint, what do these quality beef grades mean?

Prime beef is produced from young, well-fed beef cattle. It has abundant marbling (the amount of fat interspersed with lean meat),and is generally sold in restaurants and hotels. Prime roasts and steaks are excellent for dry-heat cooking such as broiling, roasting or grilling.

Choice beef is high quality but has less marbling than Prime. Choice roasts and steaks from the loin and rib will be very tender, juicy, and flavorful and are suited for dry-heat cooking. Many of the less tender cuts can also be cooked with dry heat if not overcooked. Such cuts will be most tender if braised, roasted or simmered with a small amount of liquid in a tightly covered pan.

Select beef is very uniform in quality and normally leaner than the higher grades. It is fairly tender, but, because it has less marbling, it may lack some of the juiciness and flavor of the higher grades. Only the tender cuts should be cooked with dry heat. Other cuts should be marinated before cooking or braised to obtain maximum tenderness and flavor.

Standard and Commercial grades of beef are frequently sold as ungraded or as store brand meat. Utility, Cutter, and Canner grades of beef are seldom, if ever, sold at retail but are used instead to make ground beef and processed products.
Recently, USDA collaborated with the United States Meat Export Federation and Colorado State University to develop an educational video about the beef grading process. This video provides a comprehensive overview of the beef grading system – from farm to table.

Chapter 3

Purchasing and Transporting

While color may be of interest to you, your selection should first be for carcass traits and docility. A reputable cattle buyer can help you in this area. Or, perhaps you have a neighbor, friend or acquaintance who is in the cattle raising business or knows someone who is. When you go to look for your two steers or however many you are buying, be sure to give the buyer or provider all the details of what you want and what your expected outcome or goal is. You also want to emphasize health and wellbeing of the calves. The Beef Quality Assurance Program says the calves must have two sets of vaccinations; that is, the first set and a booster at 14 to 21 days. It also states that the calves must be weaned from the cow at least 45 days. If these two conditions are met and if they are "bunk broke," the calves should be thrifty and ready to go for you.

I may be getting ahead of myself or assuming you are familiar with what you need to start with. I will give you a check off list to consider.

Determine what sex you want to buy i.e. steer or heifer.

Determine the age you want to buy. This depends on time of year and available grass or winter pasture. My preference would be a 7 to 9-month-old calf that has been weaned several days (45).

Determine the breed or crossbred, and color (optional) you prefer. Find the conformation which needs to be beefy. That will be a medium to large frame with a muscle score of 1.

Check market prices

Contact farmer, rancher, cattle buyer or friend.

Transport to your pen.

Feed and care for your steers daily or even two or three times per day.

A similar check list can be found in the back of the book. You may want to remove it from the book and post it in the barn or on your refrigerator.

One and three have already been discussed. Moving to 2 on the list, there are various classes of cattle beyond sex and breed. A weaning calf is one coming right off the cow, being weaned from nursing. The weaning calf can weigh 450 to 600 pounds. There will be a lot of bawling for the cow, walking fences and stress because that calf is not on his own yet. For a calf to settle down normally takes 14 days, for the most part. To get completely away from being dependent on milk and mothering, you'll need 45 days. A stocker is one that is ready to go to grass or wheat pasture. This calf needs some grow time to build the frame or skeletal structure.

The calf is too young for the fattening stage and needs a bit higher percent protein in the diet. This calf can weigh 500 to 600 pounds and will consume about 3 to 3 ½ percent of his body weight in dry matter per day. Grass and wheat are not dry as consumed by the calf, so he could actually be consuming 40 or more pounds of grass and sometimes 60 pounds of wheat to get the 17.5 pounds of dry matter he needs. A feeder calf is ready to be placed on a high concentrate, high energy diet that will result in stored energy and other nutrients needed in your prime beef. A feeder typically weighs 800 to 850 pounds. A finished steer is one that is ripe and ready to harvest. This steer normally weighs about 1250 to 1450 pounds.

The finished steer should show signs of fat. The primary indicators are the brisket, flanks, and pin bones. If a steer is carrying about the right amount of finish, he will be poned on each side of the tail head. Pone fat is a good indicator of quality grade for ready to harvest cattle. Some people like the pones to be the size of a softball to be considered Choice or Prime Quality.

By the way, if you have trouble with some of the terms used in this text, please go on the internet to find terms of beef cattle. A search or two will help when you do not understand the terminology. I have included a mini glossary at the end of this book also. And, don't feel bad, we all had to learn sometime. Also, remember that fat is a good term in this book. Harvest and consumption are good terms as well. In addition, medium rare is a term I like myself.

The current markets can often be found in a local

paper or an internet report, especially if you have an auction site close by. Okcwest.com has a USDA report for stockers, feeders, cows and bulls each week on their website. This is a central Oklahoma market. You will need to study one near you. The USDA report for that auction will give the numbers of cattle sold and compare them to a week ago and a year ago. It will also indicate if prices have shown a change up or down in past week and a bit more information. Then, cattle will be presented according to frame size and muscle score. An example is STEERS: Medium and Large Frame 1, which says they are medium to large frame with a muscle score of 1 or best muscle score. Once you locate this, you need to find the weight range you are looking for. Here you may see the number of head in a group, the weight range of that group, average weight within the group, a price range and/or average price. The price will be per cwt (100 #) per actual weight. All of these things give you an idea of about what your calves should cost.

An example might be as follows: Steers Medium to Large 1, 12 hd, 453-483, Ave 468 178.00 – 187.00 Ave 182.36. You could possibly purchase from a farmer or rancher a couple of calves weighing about 465 pounds for $182.36 per hundred weight or $1.8236 per pound. Calculated out, that would be close to $850.00 per head on today's market in central Oklahoma. We will talk more about total costs later on. I highly recommend you do not go to an auction and purchase on your own unless you attend several times. Watch and listen intently to all the transactions. A reputable buyer would be worth the cost of a few dollars to do your buying for you like this. Perhaps you know someone that raises some nice cattle and can purchase from them. Try to find already weaned calves that have had vaccinations.

As for transportation, oftentimes you can hire someone to haul them on a trailer to your place. If you plan to have a few calves to graze your crops or move some around, you would want to consider purchasing your own stock trailer. Be sure to get one that is solid and has a good under carriage. Keep it in good condition, too. Once the calves arrive to your place, holding them in a pen overnight is wise, and then let them see all the fences in the daylight. Hopefully nothing will spook them during the first few days on your place. Keep them fed and watered, for sure. When they first arrive, some hay usually entices them to stop, smell, and eat some. They will need some grain as soon as they can, also for quick energy to get over the stress of travel and any other stress they may have incurred.

Chapter 4.

Handling Cattle

Handling is the way you move around your cattle when checking water, feeding or just walking through. It also means how you move them or make them go where you want. In addition, it is the method of doctoring, vaccinating or any care of the animals. Handle with care, tender loving care. If you ever have a chance to view the movie, Temple Grandin, please do. This was a documentary made for HBO but is out on DVD. Dr. Temple Grandin was a professor at Colorado State University. She is best known for overcoming the disability of autism to earn a PhD in Animal Science and has written many books on handling livestock. She has been employed by many packinghouses and ranches to design facilities for handling and working livestock. She is a very interesting person, if you ever get to meet her in person. If not, you can go to her website, www. grandin.com, and find many ideas for your own acreage or pen designs. Keep in mind the number of animals you might have, because most of her work is for large groups of animals. You will want to gain ideas but make them fit your operation.

The next thing you might want to do is obtain one of her books on cattle. She explains the nature and movement of cattle and how you as a handler can respond or cause a reaction. One of the ideas from her that has stuck with me is "Cattle like to turn." Keeping this in mind has helped me design pens and alleys with respect to the points of vision cattle have and the invasion of their space to make them move in certain directions. The Beef Quality Assurance, BQA, gives a diagram of cattle movement and handler position, and I have shown it below:

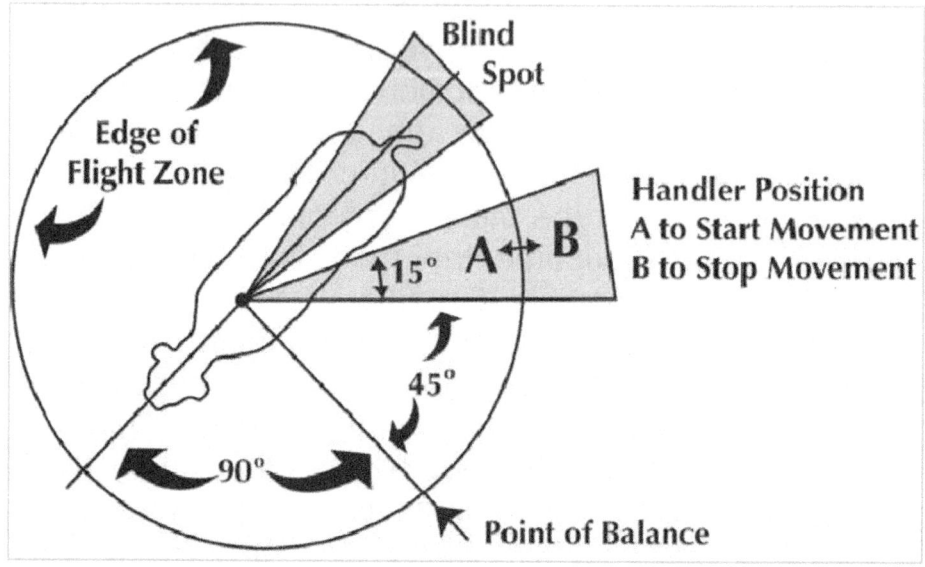

Be easy and quiet around your cattle. They will see you more than hear you and will react according to your position rather than loud voice. Do not make rapid movements of hands and arms and learn what to expect from them. They will show you how they will respond to certain and various stimuli from you. In addition, they will get used to you and whoever attends to their needs. A new person can agitate them, just by being a new person.

One year, while attending the National Western Livestock Show in Denver, I purchased a cheap sorting stick. I think it is about 3/8 inch in diameter and 4 feet in length. It is made of fiberglass, so it is light in weight. The problem with the fiberglass is splinters, though, so I taped mine using a white duct tape. One day, there were 60 calves on 160 acres, and the owner wanted them off because of the rain, so I walked out there with patience and my stick and was able to get all the calves through the gate and into the holding area with no trouble at all. I moved slow and held the stick up a bit as a guide only. Consequently, I use my stick for all my cattle work, and now I look forward to going back to Denver and maybe getting another one. I never hit the cattle with the stick. If I ever touch one with it, it is a gentle, rubbing touch that tends to make them feel good. You cannot beat or whip your cattle and expect them to have a good day, won't work. A sorting stick is primarily for the cattle to see as a guide to invading their space at a proper position. It expands your presence, if that makes sense. It is used as an aid.

A show stick can be useful in settling the animals down. If you have them in a small solid pen, you can use a show stick to rub and scratch on them. They may not like it at first but soon will desire the scratching, loving touch. When you get this mastered, you might be called a "cow whisperer." I have watched a few people work on show calves this way, and they really settle down quick. When I say quick, I mean in a few short days or maybe a few short hours but not minutes or seconds, of course. You must have patience and let the cattle figure everything out, and it works. Once you get them used to this, you could start haltering them and leading them around some or at least brush them. Many feeders and beef producers are not going to this much trouble. They may only feed, water and clean stalls for their beef production. My theory is, when they feel good and seem calm and loved, they will produce better and have better days. Plus, you won't be chasing them all over the country trying to catch them.

One other thing I might mention, when you get your calves, become familiar with their peculiarities. Some cattle like to be petted on the head while others really become obnoxious and ornery because of it. If you have a calf that starts to get mean because of something you do, do something different, quickly! You do not want to tease the animals at all in any form or fashion. That will cause you more headaches later, and neither you nor the calf will have a good day.

Chapter 5

Feeding and Nutrition

There are six basic nutrients your steers need. They are water, protein, carbohydrates, fats, minerals and vitamins. Some nutrition books may not recognize water as a basic nutrient, because it has no carbon. It is made up of oxygen and hydrogen only. However, I want to make sure it is included, because your animal's body may very well be made up of 75 % water. Free access to clean water is a must in growing your own beef. Even in wintertime, you will need to break ice or have a heated water source for the calves. Do not expect them to break the ice for themselves; that won't happen. In the summer months, the tanks can grow algae, and that should be cleared occasionally, perhaps even using Clorox to clean the tank. Fresh clean water is good stuff every day of the year. Each calf can consume 10 gallons or more per day.

Protein is important during the entire program, most of all during the growing stage from 450 pounds to 850 pounds, approximately. A tag on a bag of feed will indicate a percent protein. That represents the percent of protein found inside the bag. What you want to know is how much protein goes into your calf and a good estimate of how much he can use during metabolism. Your steers will need 12 to 14 percent protein in their diet. If a calf eats 18 to 20 pounds of feed per day, about 2.34 pounds needs to be protein. Let's say, he eats 9 pounds of hay, which is 8 %

protein and 10 pounds of a ration from a sack which is 14 % protein per day. He would get .72 pounds of protein from the hay and 1.4 pounds of protein from the ration for a total of 2.12 pounds of protein. He still needs a little bit but you will be increasing the ration as he gets used to it. This is just an example calculation. You may want to use pasture instead of hay or feed hay free choice (let the calf eat as much as he wants). Either practice will allow your calf to gain a bit more protein as he needs it.

Carbohydrates are the major source of energy. When you get into the finishing stage of growing this beef, you want your calves to store excess energy, so energy will be more important in the finishing stage where protein is more important in the growing stage. Carbohydrates include crude fibers, sugars and starches. They are found in the leaves, stems and grains of the feed source. Of course, the higher levels of sugars and starches are going to be found in the grains mixed into the ration. The crude fibers will be from the grain skins but mainly from the hay or pasture forages.

Fats from seed oils or any other source can provide 2.25 times as much energy as carbohydrates per pound fed. However, you cannot feed more fats without potentially causing other problems. Just allow your calf to consume the mixed ration, and he should get enough fats from that.

Minerals are important for all body functions. The ration and roughages fed should provide many minerals but oftentimes not all of the needed ones.

You can provide mineral as free choice in loose form or a block sometimes. The mineral mix should be what is needed for your area or what is lacking in the ration. Many of the mineral mixes have salt in them, also. Even so, you may want to allow free choice salt in the form of a block, and a trace mineral block will insure the steers getting the trace minerals as well. You can check with your local county agent to see what minerals and salt blocks are recommended for your area.

Vitamins are very important for major body functions. Vitamins A, D, E and K are fat soluble and can be stored by the animal for future use. All the others are water soluble and any excess will be excreted in the urine. Be careful about feeding vitamin supplements, because vitamins are only needed in minute amounts, are very expensive, and cattle can synthesize some of them.

A diet or ration is the amount of feed a calf will consume in a 24-hour period. If it is balanced, it will meet all his nutritional needs for that day. His nutritional needs will vary according to his age, weight and stage of production. You will need to adjust his diet accordingly. All adjustments to a diet should be gradual.

Roughages are high in fiber and somewhat low in TDN *(Total Digestible Nutrients)*. Roughages include pasture forages, hays and seed hulls among other feedstuffs. In a diet, roughages can provide a "scratch factor" to stimulate the rumen activity. I know several scientists who have used 100% concentrate diets on finishing cattle. An all concentrate diet can be used but I would highly recommend you use roughages during the entire production of your beef.

Concentrates are feeds that are typically high in TDN and low in fiber. TDN is a measure of energy. High concentrate rations are called high energy rations or finishing rations. Concentrates include cereal grains and by products.

One thing to remember is you cannot take a calf from a high roughage diet to a high concentrate diet overnight, but you could go from a high concentrate diet to a high roughage diet overnight. To start your new calves off right, not knowing what they came from, you need to provide plenty of high-quality dry roughage in the pen. Once they get used to the surroundings and the hay, you can turn them out and/or begin feeding some concentrates in the form of a ration mix. Start out with about 1 to 1.5 pounds of mix ration per hundred pounds of body weight. If your calf weighs 500 pounds, he could receive 5 to 7 pounds of mix ration. Keep this in mind during the entire growing stage, increasing the total amount as the calf increases in body weight. The calf should gain about 2 pounds per day. About 5 months of growing stage is about right and then 5 months of finishing stage gaining 3 pounds per day.

The following is a step by step suggestion or example for you to get started. Keep in mind each calf is a bit different, and the area in which you live may have different feeds available.

On the first day, have plenty of dry hay available for the calves in a pen. You will want to hold them in a pen for a while before turning them out on pasture or more space. Give them a scoop of grain mix (concentrate) in a feed bunk and see if they take to it. Allow water to be free choice, also. You should have a small amount of mineral supplement and a block of trace mineral salt in the pen, too.

On the second day, feed more hay, if needed; check water, and feed more concentrate if they ate the first day's scoop. By the way, a scoop is about the same as a 3-pound coffee can. Plastic scoops can be purchased at any feed store, and a coffee can could be saved by a coffee drinker for you.

If the calves were weaned as they should be, they should be settling down a bunch by the third day or so. You can start thinking about allowing them to go out to pasture soon but continue supplying the hay and concentrate. On the third day, increase the concentrate by ½ scoop, if they are eating it well.

On the fourth day, increase the concentrate to 2 scoops, one scoop per head, and if they are coming to the bunk when you place the grain mix in it, you can start calculating about how much grain mix you want to feed them during the growing stage. A calf will consume 3 to 3 ½ percent of his body weight in feed per day, and that may be on a dry basis rather than as fed, but let's start there.

During the growing stage, we want to feed about ½ the feed as pasture or hay. We could even use a 60

few nice steers on winter wheat pasture in Central Oklahoma

or 65% basis for the pasture or hay, if the pasture is lush and hay good quality. A round bale of hay would be good to use, and I would not be afraid to use a good grassy alfalfa bale that way. That would allow the calf to consume free choice, and the ratio of forage to grain may vary because of that, but don't worry about that. If your calf weighs 450 to start, he will consume about 13.5 to 15.8 pounds per day. One half of that is about 6.75 to almost 8 pounds of concentrate per day.

Two scoops will be close. If they eat their two scoops <u>pretty quickly</u>, you can increase it slightly, and if they eat well but leave some each feeding, cut back some.

From about the 6th day to the 60th day, you can be feeding about 2 scoops per head per day with all the hay or pasture they can eat. They should be getting very gentle in this time frame and be gaining some flesh. At 60 days, they may weigh 540 to 560 pounds. By observing their weight increase and the way they consume their feed, you may be ready to increase the amount or continue as is for a few more days.

From 60 to 110 days, you can continue the pasture, hay free choice, and increase grain to 3 scoops per day, maybe 4 scoops, 2 in the morning and 2 in the evening. Once a day feeding is kind of okay, but feeding twice per day is best, and you might consider even 3 or 4 times per day, especially when you get into the finishing stage.

Continue with the gradual increase of grain mix every 20 days or so until 180 days, at which time you will be ready to start the finishing stage.

Now, your calves should weigh 800 pounds and be ready to start a high concentrate diet. At this weight, they should be consuming at least 12 pounds of concentrate per day. Within 21 days, using a gradual increase of grain, you need to get them to 18 to 22 pounds of concentrate per day. Eight pounds increase divided by 21 days equals an increase of less than a half pound per day. If you want to increase a half pound per day for 16 to 18 days, that is okay, also. Continue free choice hay and pasture, too.

During the next 150 days, do not make any drastic

changes in feed at all. Keep a steady diet of 20 pounds, plus a steady increase up to 26 to 30 pounds of concentrate as the calves become more finished. Yes, they may consume 26 to 30 pounds of a mixed grain diet towards the end of finishing. We will talk about the costs later, and we have already discussed scooping manure (sorry, but it has to be done). After 150 to 180 days like this, your calves should weigh close to 1450 pounds, and be ripe and ready to harvest. We will discuss the processing more in detail but be sure to have a date set with your local processor before starting the finishing stage. That date should be 160 to 180 days after the finishing ration start day.

You can purchase your feed at a local feedstore in 50# bags or have a ration mixed at a local elevator that does grinding and mixing. The latter may be in bulk form or could be sacked depending on the business itself. You do want a method of storing your feed whether in sacks or bulk, and you want to keep it as fresh as possible. If you are feeding only two calves, you may want to limit the amount of your purchase to what they will consume in a two to four-week period. Your storage should be away from rodents, birds and other varmints that may want to rob.

If you decide to formulate your own rations, you can include roughage in the mix oftentimes. This can be useful if you do not have pasture or access to hay free choice feeding. Listed below are a few sample rations as suggestions only that may help you get an idea of what ingredients you can choose for your rations. You will need to consider your calves in three stages of feeding. They are growing, intermediate and finishing. The growing ration needs to be about 50% roughage and 50% concentrate. The intermediate ration should be the shortest time period and used as an adjustment between the growing and finishing diets. The intermediate ration may vary from 35% roughage to 25% roughage, with corresponding percentages of concentrates. The intermediate ration should be for approximately 4 to 6 weeks. The finishing diet should be 20% roughage to 10% roughage, and I would recommend 15%. Once you get to the finishing stage, stay with it until harvest. Only increase the daily amount as needed. Your percentage of each ingredient should remain the same.

Your ration may be bought by the bag or you could have feed ground and mixed in bulk form. If you have your own ration mixed, you may be able to get them to bag it, too. Also, this way you could select a roughage that fits your program. You can either have hay ground, to mix in ration or purchase cottonseed hulls or some other roughage from the elevator. In addition to these advantages, you can put the ration into a self-feeder and allow your calves to have the diet free choice. That would be like a smorgasbord all day and night for the calves. There could be issues with a self-feeder plan.

The calves may not become accustomed to the human touch, they may eat too much, they may bloat, you might become too reliant on the self-feeder and not check it often enough to make sure the feed is flowing, or it may run out quicker than your estimate. However, with some management, a self-feeder can be nice. This needs to be your choice, and according to your needs.

A good friend of mine used to purchase show steers at the American Royal Premium Sale and put them on a self-feeder for another 45 days before he would harvest them. Those show steers were already finished but he wanted to take them to the "next level," which was prime plus. This type of beef production is very expensive because of the size of the show steers he purchased, but he certainly wanted the over finished beef in his freezer.

Another note is that Purina has a ration that would limit intake by the steers and increase the number of times they would access the self-feeder. Now, this type of ration would work well for someone needing or wanting to use a self-feeder. A phone call to a Purina dealer or specialist might not hurt.

Today, nutritionists seem to know more about the nutrient requirements of cattle than they do for humans. I guess it depends on who you talk to. Nutrient requirements of steers vary with age, size and purpose. At 400 to 500 pounds, your steers will need about 12% of their diet as protein. They will also need about *75% Total Digestible Nutrients (TDN)*. At this stage, the steers need to grow, form skeletal tissue and muscle. Later they will need to be on a finishing diet. In addition to protein and TDN, the steers will need minerals and vitamins but not in excess.

Just because your feed tag reads 12% protein does not mean your steer is getting the 12% protein he needs. Percentages can be misleading at times. Politicians like to use percentages often; however, that is another story. The requirement of 12% means he should get 12% of his total diet or daily feed in protein. If he consumes 15 pounds of feed per day, 1.8 pounds needs to be protein. That is 15 pounds times .12 equals 1.8 pounds. You can make the same type of calculation for the TDN. That is for the growing steer calf.

When your steer gets to 800 to 900 pounds of body weight, he will need only 10 to 11% protein but will require 70% TDN. The requirement here means more energy and less protein on a percentage basis. So, if he consumes 28 pounds of feed daily, he will need 2.8 pounds of protein. That is 28 pounds times .10 equals 2.8 pounds of protein. His ration at this time will have more grain, less protein supplement and less roughage. He is now into a finishing diet.

Sample   *rations*

Receiving ration:

     Can be medicated but requires a veterinarian prescription.

     Rolled corn 35%
     Crimped Oats 45%
     Protein supplement 15%
     Salt/mineral 1%
     Molasses 4%
     Total 100

Feed 3 or 4 pounds per feeding and gradually increase each week. Have free choice hay and/or grass. You can have your hay ground and mixed into ration or use another roughage such as cottonseed hulls if available.

The above diet is about 13% Protein by itself and with the hay is about 12.4% protein. The diet is 61% TDN without the hay and almost 60% with the hay. The ration is just an example.
An intermediate ration can be the same as above by increasing the concentrate and reducing the hay a bit.

A finishing ration needs to be more 85% concentrate and 15 or less % roughage. The concentrate could be 90% with roughage at 10%. The finishing ration needs to be high in energy, low in fiber and about 10 or 11% protein. You can use a 12 or 14% bagged ration and lower the protein a bit with a lower protein roughage, such as cottonseed hulls or some grass hays.

The finishing diet needs to include plenty of corn. A ration I heard recently for finishing cattle includes 1/3 each of corn, oats and barley. This ration would be about 10.25 percent protein. I would recommend adding a mineral/vitamin supplement and feed a salt block free choice. If you decide to add a bit of molasses to sweeten the ration or to bind the grains together, I would recommend a protein supplement of soybean meal. Perhaps your ration could look like this in percentages:

Rolled Corn	30%
Crimped Oats	30%
Rolled Barley	30%

Dry Molasses   4%
Soybean Meal   5%
Min/Vit   as recommended by product maker, app. 1%

I would feed 4 or 5 pounds of hay per day unless your calves are on grass pasture. This might be called a *"scratch factor"*.

M.E. Ensminger once wrote a guide for feeding show steers that would work just as well for someone feeding a steer for their own beef. His rules or guide went something like this:

1. Use care in getting animals on feed. Do it slowly and avoid drastic changes all of a sudden.
2. Feed a balanced ration.
3. Use a variety within the ration.
4. ***Do not overfeed and do not underfeed. ***
5. Keep feed box clean and out of reach from animals.
6. Use palatable feeds.
7. Provide correct amount of bulk.
8. Do not feed damaged or moldy feeds.
9. Use prepared grains. (processed grains such as crimped   corn, rolled oats, etc)
10. Feed regularly.
11. Provide salt and minerals.
12. Keep animals quiet and content. (Never let them have a bad day.)
13. Provide adequate space.
14. Avoid scouring.
15. Avoid sudden water or feed changes.

Chapter 6

Health and General Care

Organic foods are still a sought-after product for some people. In some ways, we are getting past this fad but, we still can find stores with varying organic products. This is all well and good, but please, do not let this acronym keep you from vaccinating your calves. When you receive them, they should have had an initial vaccination of a 7-way *Clostridial, perfringens, IBR, BVD and PI3.* When you receive your calves, they should already have these, but it is cheap insurance to do it again. You can do this yourself or have a veterinarian do it.

Deworming and fly and tick control is accomplished with a *pouron* or an injectable insecticide. There are various generic products to do this, and they are over the counter products. One application when you receive the calves and one when you move into the finishing stage should be enough. You can repeat if it looks like you should.

The duration of fly and tick control with a *pour on* product is usually 30 days. Because of the short duration of effectiveness, you will need other forms of control. Spraying with insecticides around barns and pens can help. People also use fly traps, zappers, and dusts. There are some loose mineral products that contain insecticides that work on the larvae in the manure. Fly control is always an issue with livestock, and you may have ideas of your own to fight this battle.

Bloat is a gaseous entrapment in the rumen of the animal's stomach. Bloat kills, sometimes. The entrapment can be recognized on the left side of the animal between the ribs and hook bone. It will look like a basketball or larger on the left side. Some grasses, weeds, and rations may cause bloat. If you think you may have a problem with bloat because of your pasture or feed, you should use a bloat guard block, keep some *poloxalene bloat guard* on hand, and be prepared to use a hose to allow the gas to escape. Yes, a hose the size of a garden hose can be inserted into the rumen through the mouth and esophagus of the steer. The trapped gas will pass through the tube and be released. If you do use a tube, be sure the tube is placed towards the upper side of mouth and make sure the steer swallows it. Passing the tube into the air passage will not work; it has to be swallowed to the rumen of the steer. Winter wheat pasture can often cause a frothy bloat to occur. Normally, a tube or hose will not help on a frothy bloat.

Overeating can be a disaster in finishing calves. The perfringens vaccine helps with enterotoxemia, but an animal may find a time to consume more than they should. If a gate is left open, the calves will find it. And, if they can get to a feed storage, they will most likely over consume. This can cause laminitis, lactic acidosis, bloat, diarrhea and sometimes death. Prevention is the best way to control this. Keep gates shut and feed stored where no animals can access it. If an animal overeats, feed plenty of dry grass hay and no grain for a bit, use mineral oil drench, and maybe even a form of pepto to slick up the gut so the animal can void the excess. Make sure it always has access to water, too. Water is very important in all instances.

The rumen of the calf has bacteria which help digest certain feedstuffs. These bacteria can be *amylolytic* or *cellulolytic*. The amylolytic bacteria help digest starches and sugars from the grain fed, while the cellulolytic help digest roughages. Bacterial activity in the rumen is a must for your calves, and upon arrival, you may want to give them a probiotic inoculation of bacteria. This is in a paste form and can be administered orally. If for some reason, your calf goes off feed for a while and does not want to eat normally, you can try the oral probiotic again and see if that helps.

Unhealthy calves normally hold their head low, ears are drooped, and eyes may have mucous. They are lethargic in movement and may not want to get up or come to feed. Early detection is the key to survival, but we as owners may be nervous and try to see something that is not really there. A snotty nose may just be the weather, but it could be indicative of illness too. A second opinion can be helpful, if you know someone in the business. Even the farmer you purchased the calves from could help. Perhaps you could ask a county extension person, and a trip to the veterinarian is not out of line, if you do have a problem with the health of your animal.

Do not let any of this chapter scare you away from raising your own beef. These are just a few things that could happen in the course of doing so.

Normally, if you keep your cattle fed, watered and having a good day, they should stay very healthy. Even though I talked about over eating and the ills thereof, please do not under feed them, either. Watch them eat. Make sure they can get the food and swallow it. Make sure they drink water. I have been told to give them 20 or 30 minutes to finish each feeding. If they consume it quicker, you can increase the amount a bit. If they leave some in the feeder, cut back a bit on next feeding.

Chapter 7

Processing to Freezer Beef

A Clunie Range Steer photo used as shared on social media by the Clunie Range, breeders of Angus cattle in Australia. This steer is market ready and ideal for anyone's freezer. He and the steer on the cover are what you should set as a goal for your personal beef. Each one is fat, finished, ripe, ready or any other term you might want to use to describe ideally suited for the freezer.

Be sure to have a processing date scheduled in advance. Most public, local processors stay booked the year round. Plan this date carefully so the calves will be finished and ready to harvest. The harvest date can be determined by estimating the number of days on feed, and the average daily gain for your steers to 1250 or 1450 pounds. You will need to know the date when you start the finishing stage to help you select the harvest date. This needs to be estimated in advance as many good processors are booked well in advance.

As a personal preference, I would select a processing place that works with beef and pork only or at least, primarily. Beef and pork can hang in a cooler together, but I would not recommend one that hangs goat, mutton or venison in the same location. I cannot say I have had any bad experience with this but have always heard you do not want to do this. Perhaps the musk glands have an odor that might be absorbed by the beef.

In addition, you can ask the processor to hang your beef longer than normal for an aged beef factor. The processor may not be able to accommodate you on aging because of the business. If this is the case, then you can age each cut prior to preparing for a meal by thawing in a refrigerator and letting it sit a couple of days. You can learn more on aging by using Google or You Tube.

When the time comes, your processor will require a cutting order from you. If you sell a half or two, the cutting order needs to come from the purchasers. A cutting order reflects the desired retail cuts for you. I have included a chart with the wholesale cuts and a listing of the normal retail cuts. When I do a cutting order for my family, I want a bone in rib roast or two, each about five or six pounds. Remember that, and you can thank me later. I also like my steaks ¾ to one inch thick, my other roasts about 3 to 4 pounds each and my hamburger in one-pound packages. I personally prefer my round steaks tenderized and may even ask for some tri tip roast as well. I am sure they will ask if you want lean hamburger, and I usually ask for medium. It is your choice. Now, you will be getting some stew meat unless you say otherwise, and you also need to decide on heart, liver and so forth.

You will receive a rib roast, your best cut, rib or rib eye steaks, t-bone steaks, porterhouse steaks, Sirloin and tenderized round steaks. There will be 7 bone roasts as well as the Pikes Peak and rump roast. And, perhaps you will choose a tri tip roast as well. You can ask to have the brisket cut in half if you desire. Half a brisket pounds, but you may prefer the whole brisket. Remember, each side of beef will have the mentioned cuts. I am sure there will be way more hamburger meat than what you want, but it will be good and easy to prepare all sorts of dishes.

Now, let's talk about percentages again. Your calf should weigh about 1350 plus pounds for harvest. His dressing percentage should be about 65 to 68%. That would yield a carcass of about 875 pounds. Of the 875 pounds, about 595 pounds will actually go into your freezer. The reason for the decrease in weight is because you will allow the processor to have the hide and offal as well as the scraps and some bone being discarded. You may also opt to have him keep the liver, heart and kidneys. All of this is part of the processing fee, in addition to a dollar amount. You could keep the liver, heart, and kidneys for yourself if you want. I want you to be aware of this shrinkage so you will not be expecting something else. The University of Tennessee Institute of

ANGUS BEEF CHART

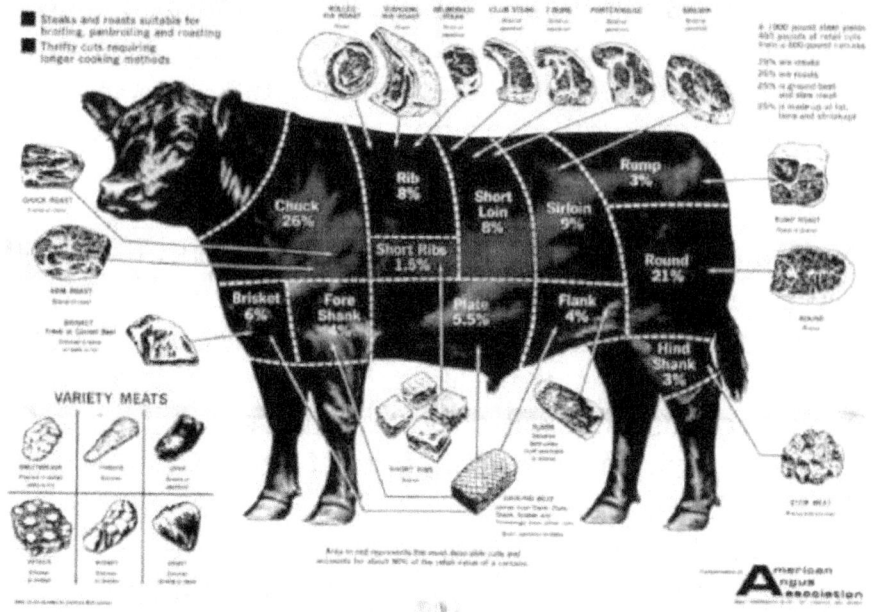

Copied with permission of the American Angus Association

Agriculture has a very good publication on this subject. It can be found on-line at:
 https://extension.tennessee.edu/publications/documents/pb1822.pdf

The processing fees will be about $600 to $800 most likely per steer. This may vary depending on your location and the processors available to you. They normally charge per pound, according to hanging weight and then add a few additional fees which would be for boning, extra trimming or tenderizing. The costs above are estimated total cost and depend on the weight and what you have done. The end product should be packaged, frozen beef for you.

Chapter 8

Let's Talk Beef

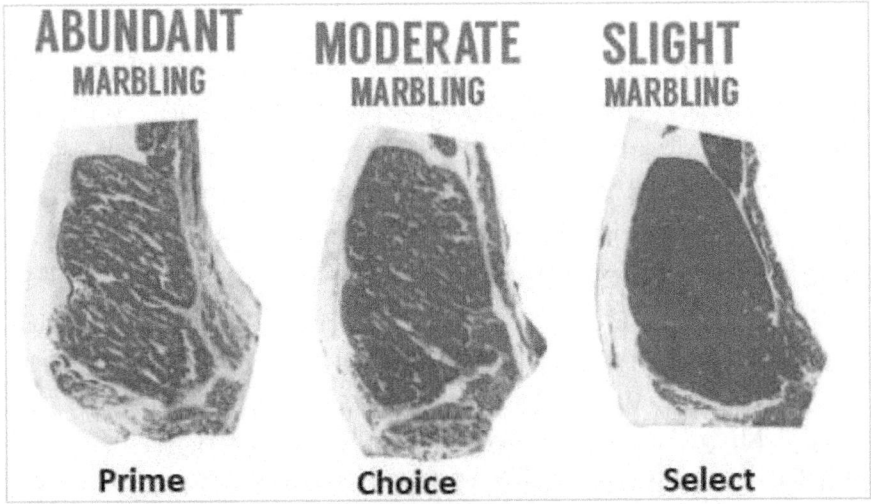

ABUNDANT MARBLING	MODERATE MARBLING	SLIGHT MARBLING
Prime	Choice	Select

Now, we want to talk about eating beef. I went to the doctor this week for a heart scan and stress test. Everything turned out very well, I might add. In the process, however, several people asked about my diet and tried to explain a lot of meat is not as good as we may have thought at one time. The gentleman who gave the stress test told me of a research that was conducted to test the vegetarian theory of remedying heart problems. During this study, the participants were allowed fish or chicken once per week and beef once per month. The results were good, of course, for the diet as such.

On the other hand, we have seen the Atkin's Diet which consists of eating all types of meat including bacon, fish and eggs. As a high fatty diet, it relies on high protein to make for weight loss. And, of course, this diet has been studied and the results are good for such a diet. Much like the vegetarian diet was found good for those studies. Hmmmmm, all these studies can make a person wonder.

Three- and one-half ounces of beef provides 26 grams of protein and 15 grams of fat along with all the good vitamins and minerals we need. It also contains about 56 grams of water. Beef is a great source for niacin, B6 and B12 vitamins. It is also a great source for phosphorus and iron. To digest protein, takes a lot of energy, thus the Atkin's Diet and a statement by one of my former professors who said, "If you want to lose weight, eat nothing but beef." Now his statement may be true; however, we would not want to eat just beef or even beef and fish as our only foods. As always, well, not always really, variety is best. Beef has a proper place and a proper amount in our diets. Another professor once said, "As long as people have the money, there will always be beef produced. He said this because the bovine is the best converter of roughages and plant products that we cannot consume to a very nutritious product that we can enjoy, beef.

The Drover's Journal is reporting a record production of Prime beef. At 9.6% of all beef production, the Prime grade is in line with the increase in Choice Grade production as well. Above Choice production is at 80%, a 20% increase from ten years ago. Since Choice and better has more marbling, the increase indicates a desire for more fat by consumers. Or, at least, more fat is acceptable. Reviewing this information indicates your steer will have a better chance of grading Choice rather than Prime, which is true. However, you should set your goal at Prime and realize Choice is not so bad either.

Thinking about this portion of the book reminds me of a goat story. I heard of a local County Agent who was known for his preparation of goat. A friend told me the County Agent knew how to make the best goat dinner ever. I asked, "How did he do that?" He said the County Agent would boil the goat for one hour, then remove it and marinate it all night in the fridge. The next day he would take the goat out and smoke it all day long, checking on it periodically to make sure there was plenty of smoke. He used an apple wood for smoke. Then, two hours before he would serve the goat, he would make sure everyone attending paid their keg fee. It was the best tasting goat ever.

Now your beef will not need to be like that at all. You will be starting with an excellent product, finished to perfection and Prime in quality. However, starting with a great product is only the start. You must have a good idea how to prepare, cook and present the final dish.

I love You Tube and sometimes just browse around looking at different subjects. On You Tube, you can find some nice recipes for beef. One of my favorites is Fool-Proof Prime Rib by Cowboy Kent Rollins. Not only does he have a good recipe, he makes the video fun to watch. By the way, my wife used his recipe for part of our holiday meal recently. It was very good.

There are many places to obtain recipes for beef dishes, using almost all the parts of the animal. Many states have a cattlemen's association, and the wives usually have a recipe book for sale. I know of some country churches that sell recipe books, too. Check them all out and find the ones that fit you best.

One thing to remember is the beef you now have is frozen and must be thawed before preparing any dish. The thawing process is up to you, but a lot of people recommend thawing in a refrigerator. This will take a bit longer than room temperature, but it will aid in an aging process if you want to age beef. Be sure to thaw your beef before cooking, especially on a charcoal grill or smoker. Take time to study some ideas on thawing frozen beef and learn about aged beef processing as well.

My wife likes to fix Prime Rib for Christmas Dinner and usually has a 4-rib roast which may weigh 8 pounds. She uses garlic, butter, salt and pepper as seasoning and occasionally thyme or rosemary. The oven will be preheated to 350 degrees and the Prime Rib is cooked to about 135 degrees internal temperature and left to rest for 20 minutes or so. No matter what recipe you use, I think the key to a rib roast is the internal temperature of 135 degrees Fahrenheit. It would be worth the price to have a good internal thermometer.

On Sundays, she fixes a rump roast or chuck roast with carrots, onions and potatoes. Saturday evenings often include charcoaling steaks or hamburgers. Oh sure, we cook bratwurst or even chicken sometimes, but beef is very important in our family's diet.

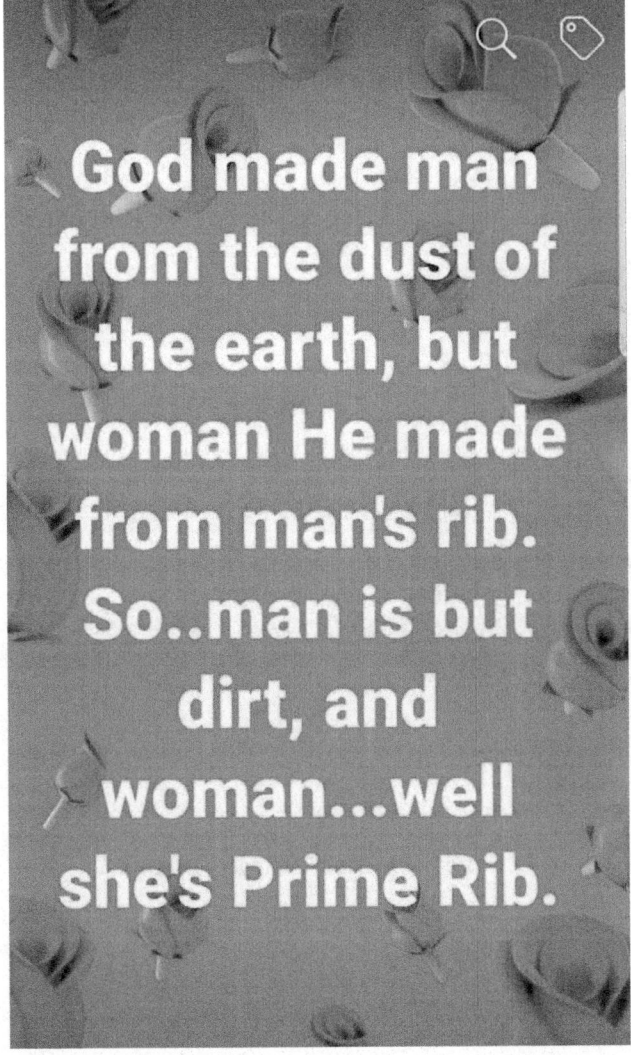

Probably a lot of truth to this, especially if she knows how to cook. Another author in our book writing class found out my title and came to class with this. She said, "I bet you will find somewhere in your book to use this." And, here it is.

RARE 👍

MEDIUM RARE ❤️

MEDIUM 😄

MEDIUM WELL 😮

WELL DONE

"I'll have mine medium rare, please".

Chapter 9

Costs and Returns

It is not going to be cheap, but it will be worth it. As I mentioned earlier, I will share some dollars and cents with you as best I can. The markets on cattle, grain and hay change often, sometimes very little change, but sometimes huge changes will occur. I want to say this, so you will realize the figures I present may not be accurate when you start your project. They, however, will give you an idea on how to pencil your project when you do start. Next is a chart indicating the costs of a two head project for producing your own beef.

Costs of Producing Two Steers for Beef
    Initial Cost 600# @ 1.50 x 2 = $1800
    Vet Medicine 30 per head $60
    Transportation $20
    Hay or pasture $300
    Feed $1440
    Processing $1000
    Total $4600

If you decide to sell three sides of beef and keep only one side for yourself, then you can recapture some of your expenses. At the above prices, you might be able to sell a side for $2.25 to $2.50 per pound hanging weight and the purchaser pay the processing fees for the side. Selling sides will not make up for all the expenses but can reduce your costs a bit.

A parent of a former student of mine fed his cattle a concentrate ration while they grazed pasture in summer or hay in winter. He fed them once per day in the evenings. He would harvest the calves when finished and sell the halves/sides at a premium as natural fed beef, because the calves were still in a large pasture. Even though these cattle received corn and other grains, they were in a natural pasture situation. This became a niche market for this person, and he soon had customers from several states and long distances to purchase his natural fed beef.

Even if your cattle are finished in a pen or dry lot, you can claim a niche practice of some sort and build a reputation for quality beef and get your beef sold. Just remember to sell it before harvest, or you may be in trouble on freezer space or other issues. The beef you produce is a degradable product and needs to be consumed accordingly.

I hate to bring this up, but stockmen know there is always a chance a live animal could die. They know this well enough that they usually calculate death loss into their projections for producing beef. The normal percentages are 2% loss on locally raised animals and about 5% on sale barn cattle. The major problem for an acreage owner is if one dies, you could have a 50% death loss. If you make sure your calves never have a bad day, you should not have an issue with death loss.

When you calculate your costs for this project, you will start gaining ideas to reduce some of the costs of production. When it is all said and done, you will be proud of what you did, and enjoy some delicious, wholesome beef. At this time, you will also make plans to improve on the previous production cycle and keep learning ways to better prepare your beef for consumption. Enjoy.

Perfection is something to always strive for. In this effort, please continue to learn more and more about the feed, cattle and preparation of beef. While you are in the process of producing your own beef, keep in mind, "Never let your calf have a bad day."

Check List

1. Decide what you want to buy, steer, heifer; breed (Angus or Angus cross) ; weight, etc.

2. Check market prices and know a close amount for purchase

3. Find a rancher, farmer or order buyer and locate the animals you need.

4.****Make sure they have all vaccinations**** worming and castrations.

5. Make sure they have been weaned a minimum of 21 days and   preferably 45 days.

6. Have your pens ready to receive them. Make sure gates and   fences are good and properly secured.

7. Have pasture fences checked and ready.

8. Check water daily and check health daily.

9. Free Choice hay for entire period is good except when you have lush pasture for them to eat. Grow your steers to 800 or 900 pounds feeding some concentrates along the way.

10. Re-worm and maintain fly and tick control.

11.Full feed equals plenty of concentrates and finishing diet until harvest date.

12.Get the finished steers to 1350 pounds or even 1450 pounds might be preferred. They need to be fat, ripe and ready to harvest.

13.Make sure you have a processing company lined up way ahead of time.

14.Decide on a cutting order, place that order with processor.

15. Keep your freezer running.

16.Thaw each cut properly before cooking, maybe even age the beef.

17.Use good recipes for all your dishes.

Cowman Terminology

7-Way Clostridial – A vaccination that includes *Clostridium chayvoei* (blackleg),
Clostridium septicum, *Clostridium sordellii* (malignant edema), *Clostridium novyi* (black disease) and three types of perfringens (*enterotoxemis*)
IBR – Infectious bovine *rhinotracheitis*
BVD – Bovine Viral Diarrhea
PI3 – Parainfluenza-3
Finish – Fat cover on a carcass; fattening an animal for harvest
Finished – Ripe and ready to be harvested
Poloxalene – A block polymer of ethylene and propylene oxicides.
It can be fed daily or as a drench when needed
Amylolytic Bacteria – Digest starches and enjoy a slightly acidic pH
Cellulolytic Bacteria – Digest cellulose and enjoy a slightly alkaline to normal pH
Castration – removing the primary sex glands usually of a bull to make him a steer
Steer – Castrated male bovine
Heifer – Young female bovine
Bull – Intact male bovine
Stocker – calf that has been weaned and ready for pasture
Feeder – calf that has grown to about 800 or 850 pounds and is ready for full feed. Ready for a high concentrate diet to be finished out
Fats – cattle that are ripe and ready to harvest. They are finished or fat
Marbling – Intramuscular fat INTRA being

emphasized here

Ribeye – The muscle exposed when a carcass is split between the 12th and 13th rib. This muscle is often measured for area and recorded in square inches.

Quality Grade – determined by the marbling and maturity of the beef

Yield Grade – Determined by the back fat measured ¾ the distance from the chine bone to end of ribeye, ribeye area, and kidney, pelvic and heart fat percentage.

Side – A side of beef is ½ of the total beef. A quarter is a term for ½ of a ½ but it has to be a fore quarter or a hind quarter and there is a major difference in price with the hindquarter being much more valuable

Organic – Growing agriculture products organically requires certain stipulations to be met. You can look up USDA Requirements for Organic Food Production

Grass Fed Beef – Beef produced by feeding only pasture and hay. The cattle gain less per day this way so it will take longer. The fat is different in color and deposition also. There is less quality in grass fed compared to grain fed

Natural fed Beef – Cattle fed grain while in a natural pasture environment. Not in intensive confinement such as a pen.

Aged Beef – This is beef product which has been held at approximately 38 degrees for a period of time, two or three days even during thawing. Beef can be aged in the carcass shape by asking a butcher to allow it to hang for a week or so before cutting into processed units. This usually does not happen because the butcher needs the rail space for more income.

Caliche - A mineral deposit of gravel, sand, and nitrates, found especially in dry areas of South America. An area of calcium carbonate formed in the soils of semiarid regions.

Poloxalene **bloat guard** – A mixture of mineral oil and *poloxalene*, sometimes called *therabloat* and may be a brand name of certain products used to let down a bloat.

Head Gate – An apparatus used to catch a bovine at the neck. It does not squeeze the neck but is tight enough to keep the head from going back and the shoulders from going forward. A head gate does not squeeze the body, it only holds the head. It is used for minor work on the animal and calving assistance. It is not good for surgeries or branding.

Stocking rate – This is the number of cattle that can be placed on a given pasture, in central Oklahoma, the stocking rate might be 1 animal unit per four acres. A field of 100 acres might have a stocking rate of 25 head as an example.

Pouron – A *pouron* insecticide or chemical is one that works systemically and is absorbed through the skin. *Ivomecterin* is an example

Niche market – Marketing a product to a few buyers. The buyers are usually few because it is not a widely desired product because of various factors.

Bunk broke – Cattle that have been trained to come to the feed bunk at feeding time are considered bunk broke.

Receiving ration – first ration to be fed. This ration could be medicated as a precaution

Beginning ration – a low concentrate high fiber diet to start the finishing process. This diet should be abandoned in a short period and the calves moved to an intermediate diet

Intermediate ration – higher concentrates in the adjustment period of changing from high roughage to high concentrate rations

Finishing ration – the final high concentrate ration usually has plenty of corn and used to finish the animals. This high concentrate diet needs to be approached gradually and not all of a sudden. This is a high energy ration, sometimes called a hot ration

Worming – using an insecticide like *ivomecterin* to kill worms in the digestive system. *Ivomec* will kill flies and ticks as well and can be poured on or injected

Scratch factor – pressure against the rumen wall stimulates rumination in the steer

Notes:

I am happy to be able to use the Horn Family on the cover of my book. Brandon is a former student of mine and his wife Brek gave me permission to use the photo with Bentley, the 1400 pound, Charolais cross steer shown by Aven who was helped during the feeding and process by her brother, Jagger.

Congratulations to Aven on her Grand Champion. Also, M. E. Ensminger used a photo of her dad, Brandon, and his Grand Champion Steer in his Feeds and Nutrition book.

I want to thank the Ft Worth Stock Show and Rodeo for the cover photo of the 2019 Grand Champion Steer shown by Aven Horn, pictured with her is Brek, Brandon, and Jagger.

I also want to thank the Callis family, Brandon, Kelly, Bradon, Kylie and Cambrie for allowing me to use their photos to illustrate parts of the text.

Many thanks to the American Angus Association and Clint Mefford. And, thanks to the USDA for the data they share.

My email is *edzweiacher@gmail.com* and I am on facebook. You can find some of my photos on flickr.com as SwissEd.

Thank you for viewing and purchasing.

Made in the USA
Middletown, DE
16 July 2021